STEVEN D.J. BAUMANN

A BRIEF LOOK AT
EARTH'S ICE AGES

A Brief Look at Earth's Ice Ages
Steven D.J. Baumann, P.G.
©2014
Midwest Institute of Geosciences and Engineering
(inside cover)

Cover Photo taken by Steven D.J. Baumann on May 27, 2014. Photo is of the outcrop containing the tillite within the 2.34 Billion year old Gowganda Formation in Ontario Canada off of Route 638. Person climbing for scale near the center.

A BRIEF LOOK AT EARTH'S ICE AGES

STEVEN D.J. BAUMANN, P.G.

Midwest Institute of Geosciences and Engineering

© 2014

www.mige-web.org

EDITED BY: **SANDRA K. DYLKA**

ISBN-13: 978-1502794680
ISBN-10: 1502794683

TABLE of CONTENTS   **PAGE**

GLOBAL ICE AND YOUR PLACE IN A DYNAMIC WORLD

If you do an internet search for "ice ages", you will get a lot of data about the glaciers that covered the Northern Hemisphere in the past 2.5 million years. This is referred to as the Quaternary glaciations. Technically, we are still in the ice age. However, earth has experienced several known periods of glaciation extending back to at least 2.5 billion years.

What is an ice age? Most of us are familiar with alpine glaciers. These are the ones that occur high in the mountains. During an ice age, continental glaciers spread out to cover the land. Continental glaciers make alpine glaciers look like an ice cube on top of a snowman. At present, only Greenland and Antarctica are covered with continental glaciers. This has not always been the case. You can't have an ice age without continental glaciers.

Ice ages are interrupted with short periods of warmth called interglacial periods. We are in one such interglacial period right now. In this time all of human civilization has been born and thrived. However, it cannot and will not last forever. That's where you come in. Human beings are unique because we are not only capable of understanding our world, but we can change it. That's what we have been recklessly doing for the past 200 years as if we can continue with our civilization indefinitely. As if some miraculous technology will be discovered in the last hour. Or we just don't care because we will be dead by then. You may not actually be dead by the time the climate shifts. Technology cannot solve every problem and the status quo is not sustainable.

Maybe you do care. Perhaps you want civilization to continue after you are gone. If we continue as we are, we will drive ourselves to extinction. Not only is the climate changing but we are consuming and abusing resources at an alarming rate. Our present civilization structure is unsustainable. The Earth only contains so much oil, copper, iron, so many trees, and only so much fertile land that we remove by building condos. Only we can change our fate. In order to understand how we do that responsibly, we need to understand our past. I'm not just talking about what is in a history book. We need to go back much further.

A key to understanding what will happen in the future is to study the ice ages. Not just the recent ones with wholly mammoths, saber toothed cats, and the humans that hunted them. We need to study older ice ages, extending back an unimaginably long time. Before your grandparents were born. Before the industrial revolution. Before there were any humans at all. Even before the dinosaurs. Yet this still isn't far enough. We have to go back before there was even life on the land or oxygen in our Atmosphere billions of years ago. The history of the Earth is not in a museum or a book. It is in the rocks. The ability to read the rocks not only tells us about the past, but the future as well.

We are currently in an ice age, but in between cycles. Ice ages are not continuous and without change from birth to death. Like everything else in nature, ice ages cycle. It helps to start with the familiar. So lets start with today and work our way back, deep into the past.

SIMPLIFIED TIME CHART of the HISTORY of EARTH

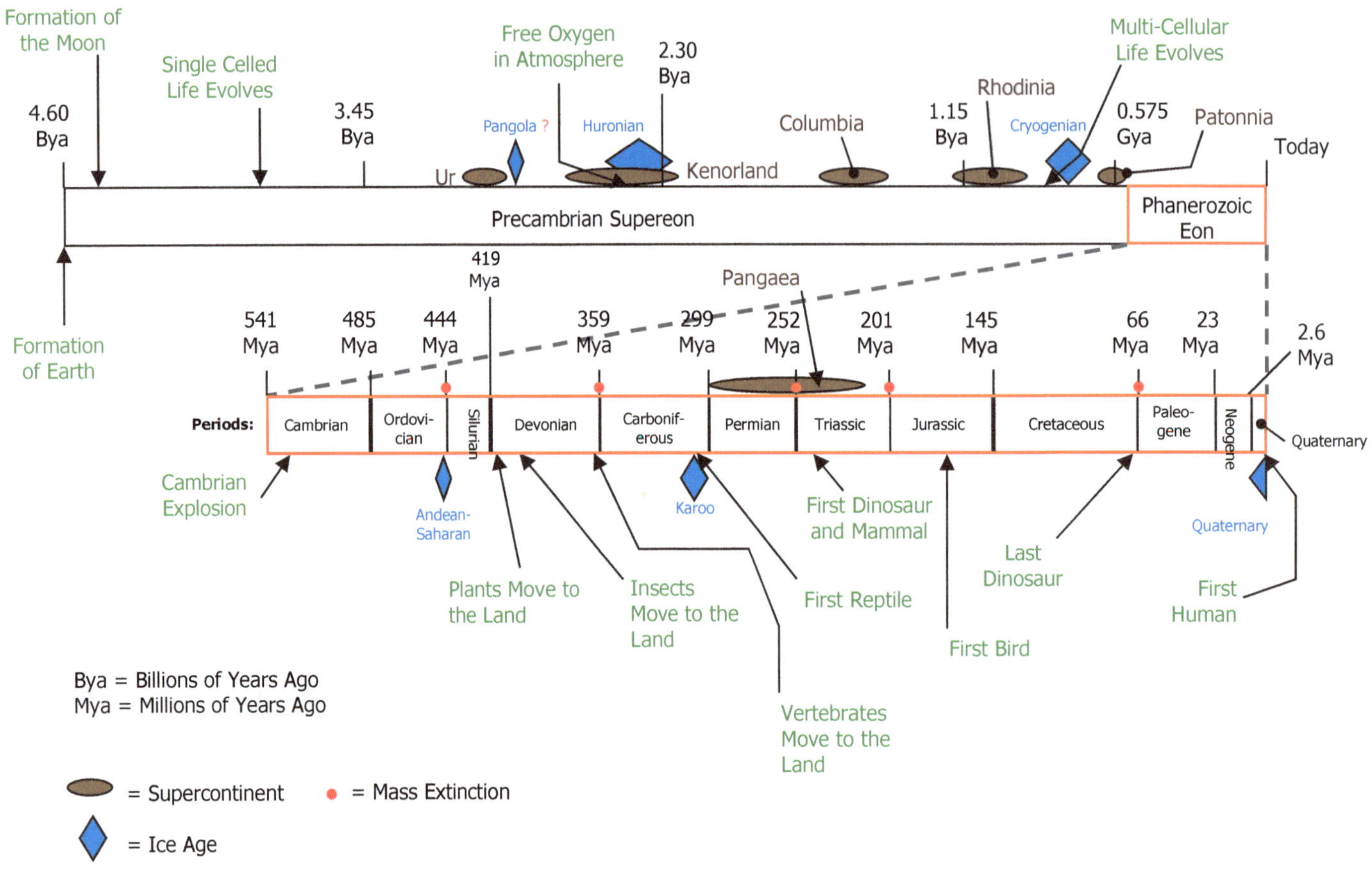

Formation of the Moon

Single Celled Life Evolves

Free Oxygen in Atmosphere

2.30 Bya

Multi-Cellular Life Evolves

Rhodinia

Columbia

1.15 Bya

Patonnia

4.60 Bya

Pangola ?

Huronian

Cryogenian

0.575 Gya

3.45 Bya

Ur

Kenorland

Today

Precambrian Supereon

Phanerozoic Eon

Formation of Earth

419 Mya

Pangaea

541 Mya | 485 Mya | 444 Mya | 359 Mya | 299 Mya | 252 Mya | 201 Mya | 145 Mya | 66 Mya | 23 Mya

2.6 Mya

Periods: Cambrian | Ordovi-cian | Silurian | Devonian | Carbonif-erous | Permian | Triassic | Jurassic | Cretaceous | Paleo-gene | Neogene

Quaternary

Cambrian Explosion

Andean-Saharan

Karoo

First Dinosaur and Mammal

Quaternary

Plants Move to the Land

Insects Move to the Land

First Reptile

Last Dinosaur

First Human

First Bird

Vertebrates Move to the Land

Bya = Billions of Years Ago
Mya = Millions of Years Ago

⬭ = Supercontinent   ● = Mass Extinction

◆ = Ice Age

COMPARING GENERIC CROSS SECTIONS OF STREAM SANDS AND DIAMICTON

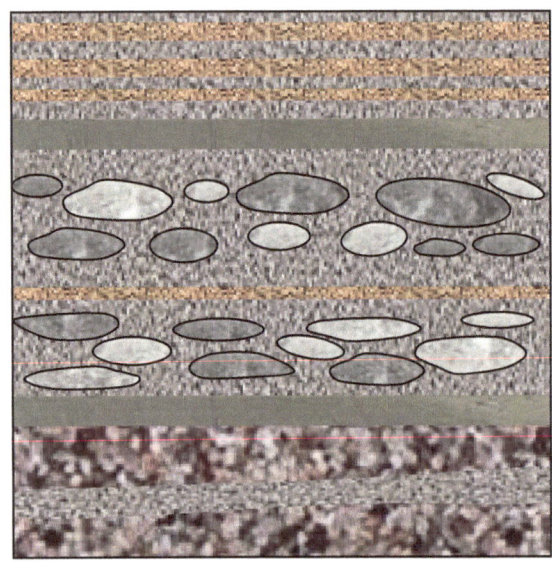

~1 Foot

Hypothetical cross section of stream sediments, showing how they are layered (bedded) and the pebbles tend to be the same size and orientation.

Hypothetical cross section of diamicton, showing how the matrix is fine grained and a consistent color. The pebbles to cobbles are a variety of orientations, sizes, shapes, and rock types.

QUATERNARY (RECENT ICE AGE)

The Quaternary glaciations are the most studied because they are recent. We have a relatively complete picture of them. The further you go back in time, the more evidence that has been erased by our changing world. Understanding the Quaternary is the key to understanding older glacial cycles. Although, the further you go back in time the more unrecognizable Earth becomes.

A Brief Look over the Last Two and a Half Million Years

An ice age occurs when ice advances from the polar regions towards the equator. The Quaternary ice ages have been restricted to the northern hemisphere, reaching as far south as 37.5° north latitude. This is more than half way to the equator!

You may ask, "if we are still in an ice age why is the ice only near the poles"? We are in an interglacial period. During an ice age, the sheets of ice grow and shrink in slightly regular cycles of about 100,000 years. I use the term "regular" loosely. There are significant deviations. Right now, we are in an interglacial period called the Holocene, which began about 9,500 to 12,000 years ago. The interglacial period began sooner at lower latitudes. Arguably, it never ended above the Arctic Circle (66.56° north latitude to the north pole). The Holocene began later in the north as the ice sheets retreated. Modern human beings have witnessed the growth and withdraw of the most recent ice age (called the Wisconsin Episode). Other humans witnessed the Wisconsin Episode and older cycles. The Neanderthals saw at least three major glacial advances.

Interglacial periods can last as long as 20,000 to 30,000 years. So we would be about halfway through the present interglacial. The last interglacial ended about 120,000 years ago and the temperature was slightly warmer than today. Areas such as Greenland and Antarctica presently contain more ice than they did during past interglacial periods. Indicating that the Earth is presently in flux. The question is will we get warmer or colder? Before we can attempt to answer this question we need to understand what drives these ice cycles.

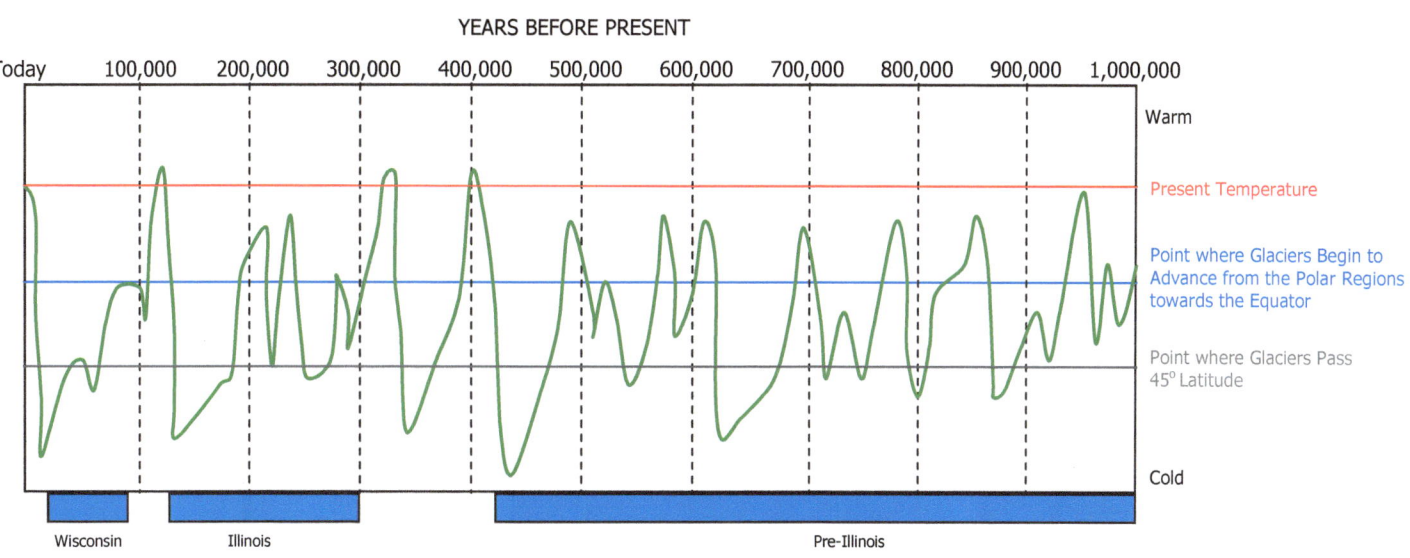

SIMPLIFIED GRAPH of RELATIVE GLOBAL TEMPERATURES
OVER the PAST 1 MILLION YEARS

What Do the Rocks Say?

The only reason we even know about ice ages is because of the deposits that they leave behind. Glaciers leave outwash (sand and gravel), lake sediment (silt and clay), loess (wind blown silt and fine sand), and diamicton (roughly equivalent to till) behind. They also leave behind distinct landforms such as moraines, kames, drumlins, and eskers. Outwash is mostly sand and gravel that is deposited in thick sheets, by fast moving water, as glaciers melt and their associated lakes drain catastrophically. Lakes such as the Great Lakes, formed where glaciers once stood. Diamicton is deposited directly by the ice, under or in front of a glacier. The hard rock form of diamicton is called tillite.

Glacial deposits are unique and easy to tell from rocks deposited in a marine or river setting. Rocks in marine and river environments tend to be layered or and exist in relatively flat beds. Deposits left directly by glaciers tend to be jumbled and unlayered, as in the case of diamicton. Outwash and lake deposits are usually bedded, but they too can be tied to glacial origins if they interbed with diamicton or are deposited near it.

Diamicton is roughly equivalent to "till". The term till has slowly been falling out of favor since the 1990's. The reason being is that till suggests a glacial origin, as where diamicton is a descriptive term. Diamicton is a jumble of clay, sand, silt, gravel, cobbles, and boulders. In most places diamicton is dominantly clay with silt and some sand. However, this isn't always the case. Sand diamicton does exist, especially in the Upper Peninsula and Ontario. However, tillite is still a widely used term, although diamictite is beginning to replace it.

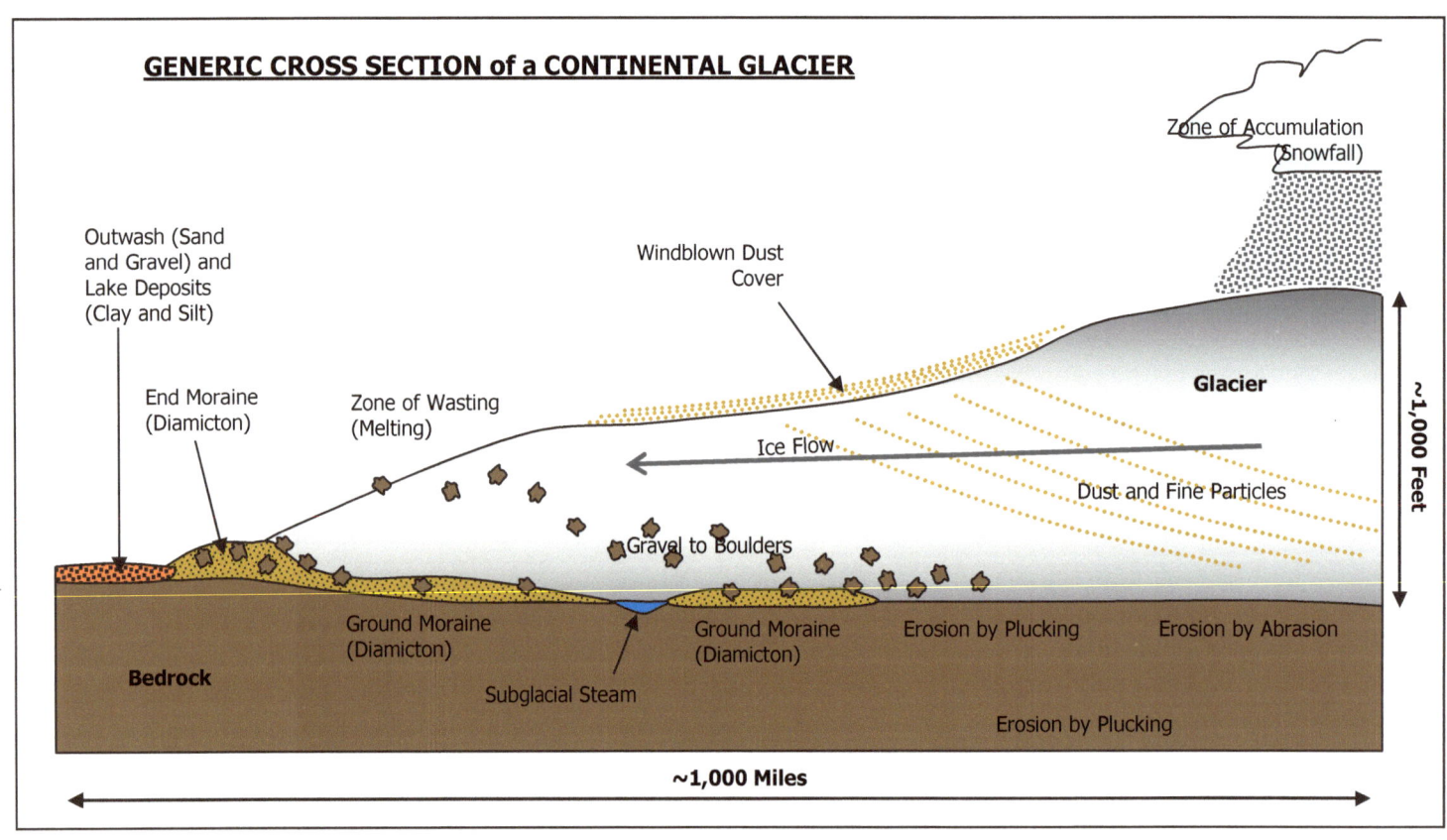

The composition isn't what makes diamicton distinct. No matter if it is dominantly clay or sand, it is jumbled together. This type of well mixed deposit is referred to as "unsorted", because unlike most deposits there are no distinct layers and variable grain size. They contain something else unique. Diamicton will often have large random cobbles and boulders that are unlike the local rock. These inclusions of large rocks can come from hundreds of miles away and appear to be "floating" within the finer sediment! Glaciers pick-up rocks as they move forward and incorporate it into their mix. Once they melt, they drop all of their load and deposit diamicton (along with other deposits). Glaciers erode the landscape as they grow. However, they do not function like giant bulldozers. Instead they pick-up and scour (process called plucking and abrasion) the rocks that they pass over and incorporate the material within. Glaciers will often leave straight groves on rock surfaces as they drag cobbles at their base, leaving behind striations on the surface. Striations can be used to determine the direction of ice flow.

Diamicton can be only several feet to a few hundred feet thick. As each new cycle of glaciation begins, they carry slightly different material with them. This variation in material can be used to differentiate different glacial cycles. The ability to differentiate glacial cycles does not tell us about when they were deposited. The good thing about the most recent glacial cycle is that they leave things behind that can be dated, such as wood fragments and pollen. During interglacial periods, soil will form called geosols. Geosols are ancient soil profiles that form when the land is ice free, and resemble modern topsoil. Glacier are rarely clean. The beautiful picturesque scene often photographed are usually of modern alpine glaciers. Glaciers are actually very dirty things, but who wants to see a photograph of a glacier that looks like mud has been poured over it? Glaciers are carrying countless tons of not only rock debris but also things like tree trunks and branches. If preserved, the tree debris can be dated as well by using carbon-14. Also, fossils in glacial lake deposits can be dated. The use of Oxygen isotopes (such as O-18 levels) locked in the modern ice sheets can also be used to determine warm and cold periods.

What Drives the Ice?

Once we can recognize glacial deposits and their cycles we can attempt to explain why they happen. There are many theories. Like most things that explain the natural world, one theory to explain it all is not practical, nor is it likely. Humans like things to be simple. Earth is a dynamic place that doesn't conform to us. That's why rivers are not straight, nor are mountains pyramid shaped. Earth conforms to the laws of nature not humanity. Although the laws of nature always apply, they can sometimes be at odds with one another. This creates a complex and dynamic environment that is the combination of many factors.

Since the Quaternary glaciers occur in relatively regular cycles of about 100,000 years, maybe something astronomical is driving them? In the early 20th Century a Serbian astronomer and geophysicist by the name of Milutin Milankovic proposed that the precession, eccentricity, and axial tilt can be used to explain climate changes. These are now called Milankovich Cycles. The orbit of Earth around the sun is stable, but not perfect. The orbit of our planet around the sun is not circular. It is an ellipse. This means that at times Earth is closer to the sun than other times. During the closest (perihelion) the Earth is about 91 million miles from the sun. At its furthest (aphelion) it is about 95 million miles from the sun. There are other variations as well. Due to contrary belief, the northern hemisphere experiences winter when the Earth is closer to the sun. This may seem like a contradiction but the Earth does not rotate about its axis at $0°$ from its orbital plane.

Earth is presently tilted at about 23.4° to its orbital plane. So during winter, even though the Earth is closer to the sun, the northern hemisphere is tilted away. This means shorter days and thus winter for the northern hemisphere. Even the tilt of our planet isn't stable. Earth's orbit varies in about 26,000 year intervals, known as precession. Precession causes the tilt in Earth's axis to change about 2.4° over time, like a spinning top. This change in orbit along with eccentricity has been used to explain the ice ages. However, this is a far from complete picture. It is likely that the Earth's orbit may cause regular cycles in the Quaternary Ice Age, but cannot be the sole cause. The biggest hitch to using astronomical events as an only explanation, is the fact that Earth does not usually have ice caps.

Scientists are human just like everyone else. Human beings like to think in terms we are familiar with. We see ice caps on Earth today and we tend to think this is the norm. It isn't. Earth has generally been a warm planet. All the combined ice ages from the formation of Earth to today may encompass a total of 100 million years. This is about 2% the entire time of Earth's history. Antarctica was the first to freeze over in recent geologic history, about 25 to 34 million years ago. This is far before the Quaternary. Areas such as the north polar regions and Greenland froze over during the Quaternary. Ice caps are more common than actual glacial episodes. Their evidence is also more obscure. It is likely that Earth has had ice caps on and off for a grand total of about 580 million years. This is roughly 12% of the entire history of our planet. Before Antarctica froze over, we have to go back 260 million years before the rocks record evidence of ice caps. This is about 20 million years before the first dinosaur appeared and just prior to the Permian Mass Extinction. Even more rare than ice caps, appears to be ice caps at both poles. As far as we can tell this is the only time in Earth's history when we have evidence for both poles freezing over. Does this mean it is the first time to happen? Not likely. When one cap is over an ocean, the evidence gets recycled relatively quickly by plate tectonics.

Plate tectonics is the ultimate driving force of our planet. It is how Earth sheds its internal heat. The process has been going on for about three billion years and will likely continue for another one billion before the Earth's internal heat can no longer drive the plates. The surface of the Earth is not fixed. It moves...a lot! Continents split apart and reassemble. Oceans grow and shrink. Mountains rise and fall. Volcanoes are born and die. The Earth is divided into 12 large plates. These plates are separated by ridges and trenches. As magma moves up through the Earth's interior and reaches the surface, the surface cracks, forming a rift. As the rift grows, plates spread apart. Since the Earth isn't growing in size, somewhere the surface has to be getting recycled back into the interior. This occurs at deep trenches in the oceans called subduction zones. This rifting and subduction consumes the rocks of the oceans. Ocean rocks are more dense and are easily subducted. Most of the ocean floor dates back to the Triassic-Jurassic boundary (about 200 million years ago), just as dinosaurs got their foothold. The oldest ocean crust is only 270 million years old in the Mediterranean Basin. The continents are less dense and thus more buoyant. They do not get recycled. As a result, the rocks can be much older. The oldest known rock is from northwestern Canada and comes from the Acasta Gneiss (a geologic unit). It is 4.31 billion years old! Most rocks on the continents are much younger. The continents are exposed and highly susceptible to erosion. In North America, the rocks generally get older the further north you travel from the Gulf of Mexico to Baffin Island. As the oceans grow and shrink, the continents sit on the plates safe from destruction into the mantle.

TECTONIC PLATES SURROUNDING NORTH AMERICA

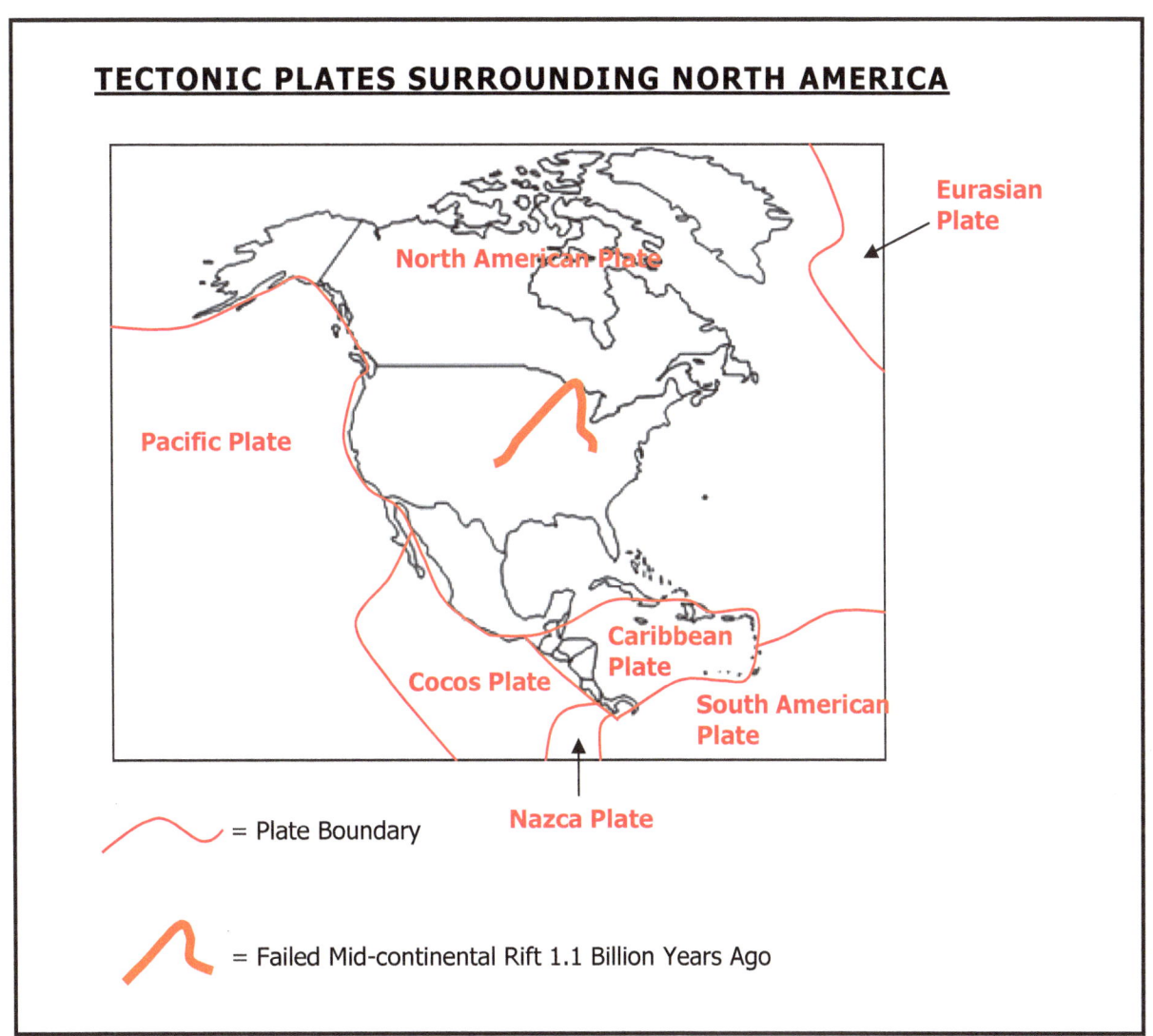

Eurasian Plate

North American Plate

Pacific Plate

Caribbean Plate

Cocos Plate

South American Plate

Nazca Plate

⌁ = Plate Boundary

⌁ = Failed Mid-continental Rift 1.1 Billion Years Ago

What does this have to do with ice ages? A whole lot! As the continents move about the globe by the ocean plates (as they spread and subduct), they occasionally cluster in one hemisphere or the other. You have probably heard of the super continent Pangaea. Pangaea existed when all the continents were together in one spot on the globe. Pangaea was centered near the equator and topography was low, so there were no glaciers. However, Pangaea is not the only supercontinent to have existed. It is just the youngest to have existed. Vendia (also known as Patonnia) existed about 600 million years ago, during a Snowball Earth Event. A Snowball Earth Event is when there was so much ice on the planet that the entire surface of the planet was nearly frozen over. Rhodinia existed about 1 billion years ago. Columbia was around about 1.7 billion years ago. Kenorland existed during the oldest known confirmed ice age about 2.6 billion years ago. Ur was around about 3 billion years ago and its arrangement is highly speculative. Vaalbara possibly existed 3.6 billion years ago, but present evidence is scarce. Some of these supercontinents like Kenorland and Vendia were clustered above or below the equator. Ice forms over land more readily than over the oceans. When these continents clustered in one hemisphere, ice grew, forming ice caps and eventually glacial periods. Although the modern continents are not all clustered at one pole or the other, they are clustered in the Northern Hemisphere. 64% of Earth's land is north of the equator, this is almost two thirds. You know this if you have ever looked at a globe from the top and compared it to the bottom. There is a lot more land up top.

So when continents group together in either the Northern or Southern Hemisphere, Earth cools. That helps explain why we have not continuous ice ages over hundreds of millions of years. Why did the ice sheets advance 2.5 million years ago? Why not when Antarctica froze over? Were the continents in different positions? The continents were located close to where they are now at the beginning of the Quaternary. However, they were not exactly as they are today. There was something that may seem small on a global scale, but when it happened it changed the surface of Earth.

The collision of North and South America had a profound change on the life and climate of Earth. When the two met about 4 million years ago and permanently connected about a million years later, things changed. Panama formed as the land rose up from the sea, and the North American Plate began to pass over the Cocos Plate forming Panama. Prior to the uniting of North and South America, the Pacific and Atlantic Oceans could mix at middle latitudes. Once Panama formed, the two could only mix in the polar regions. This changed Ocean circulation on a global scale. The Gulf Stream could no longer exit partially into the Pacific Ocean. The formation of Panama helped bring large amounts of moisture up to the cool and dry northern latitudes, increasing snowfall and encouraging the growth of ice sheets. Humanity isn't just an innocent bystander to nature.

Human civilization is helping to warm our planet. However, we are still in an ice age. The warming can flip the climate switch instantly. Past advances of the ice sheets have been quick on a geologic time-frame. The climate can flip on a dime and glaciers can be south of the 45th parallel within several thousand years. From about the year 1550AD to 1850AD, Europe experienced what has been called the Little Ice Age. During this time glaciers advanced, destroying farming at high elevations. Although overall temperatures dropped, the weather became erratic. Some winters would be bitter cold with no snow. Other winters would be somewhat mild and see snowfall that persisted until late spring. Many summers were cold and wet only to give way to early snowfall. If this trend had continued into the 21st Century, we likely would be in the beginnings of the next ice age today. The trend did not continue. World ice cover is on the retreat. Are humans to blame? Yes and no. We are unquestionably adding to the variable climate swings seen in the last 50 years. Are we the sole cause? I would not be completely honest if I blurted out, "Yes, we are!".

People tend to view time based on their life span as being typical or normal. We look around our world and assume that what we see is always how the world always has been and ever will be. Since a human life isn't even a blink of an eye when compared to the age of the Earth, we cannot just say that what we observe today is normal. Everything changes and the Earth is no exception. Just as when run your heart beats faster. If I took your pulse at the moment after a run, it would be higher than normal. Do I assume that your high pulse is the norm? No. If I take your pulse, it is a snapshot in the longer time span of your life. It is not representative of what is "normal". Your life is a tiny short snapshot in the lifespan of our planet.

Other factors that may contribute to ice ages are volcanic eruptions, unusually low carbon dioxide and water vapor in the atmosphere, the variability in the output of the sun, the distribution of forests, and changes in ocean chemistry and currents. We do not completely understand the driving forces behind climate change, but we have a pretty good idea.

The three most recent ice ages are well preserved in the rock record, especially in North America. From oldest to youngest are the Pre-Illinois, the Illinois, and the Wisconsin Episodes. Each one of these is separated by a warm period. The Pre-Illinois may actually be several ice advances and retreats that did not fully melt back to the northern polar latitudes. There is a small trend within the recent ice age. In north America the glacial record is hard to nearly impossible to chase back more than a million years.

This doesn't mean there were no glaciers. They may not have reached very far south or their evidence has been eroded by subsequent ice sheets. Starting a million years ago the ice sheets reached much further south, yet the climate wasn't as cold as it was during the most recent Wisconsin Episode. At about half a million years ago, the ice sheets stopped reaching as far south as they previously had. During the Illinois Episode the ice sheets almost reached the Ohio River (about 37.5° latitude) and covered almost 90% of Illinois at their maximum. During the Wisconsin Episode the glaciers only reached about 39.5° north latitude and covered less than half of Illinois. The seemingly perplexing thing about this is that the ice sheets made it further during warmer glacial cycles. This is due to the fact that the colder ice is, it behaves less like the cubes in your tray. Warmer ice behaves more like a fluid. Colder ice behaves more like rock. When it's colder the ice may be thicker and heavier but it doesn't move as quickly. As a result a cold glacier doesn't make it as far south as a warmer glacier.

EXTENT of QUATERNARY ICE SHEETS
On NORTH AMERICA

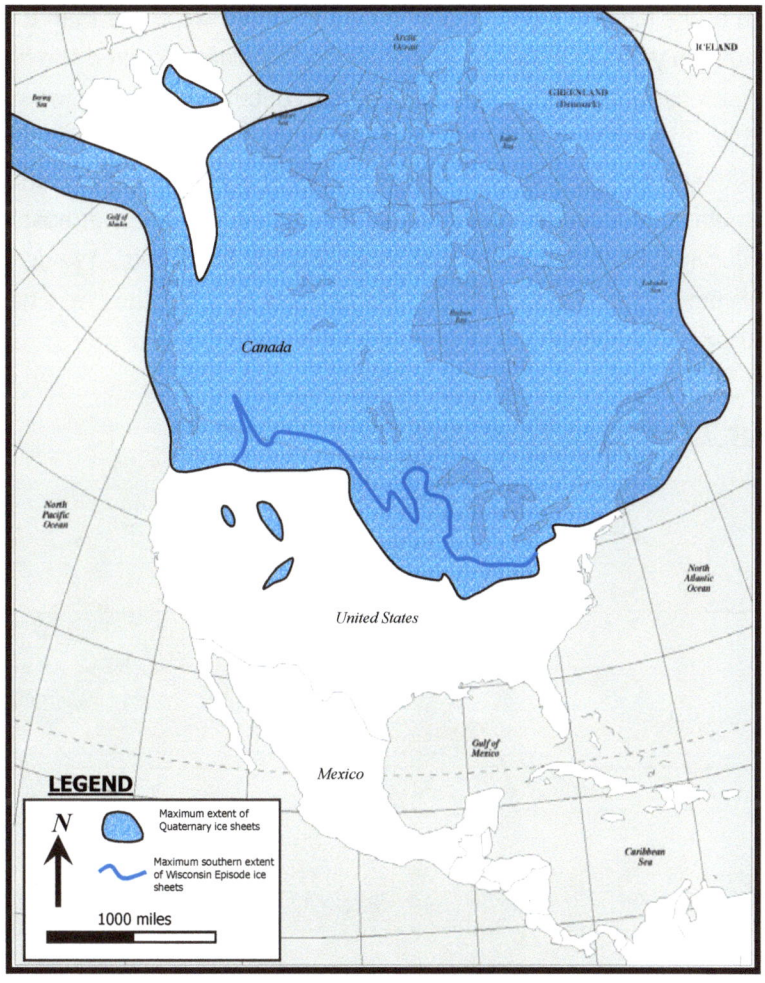

KAROO ICE AGE

Now that we understand how to recognize glacial deposits as well as what drives ice cycles, we can begin to leave the Quaternary and address older ice ages. The next youngest ice age is called the Karoo Ice Age. It is named for the Karoo Region of South Africa where the glacial diamictons were first recognized. The span of time for the Karoo is 260 to 360 million years ago. This is before the dinosaurs and before Pangaea. During this time more than half of Earth's land was south of the equator. India and South Africa were almost at the South Pole. It was during this time that ice began to form. Although ice likely covered the south polar region for this entire time, the actual ice ages were more limited.

What Was the Extent of the Karoo?

It's now known that the ice sheets advanced twice between 299 and 318 million years ago. That is a 19 million year span, more than 7 times longer than the Quaternary Ice Age. At this time the ice age was restricted to the Southern Hemisphere. The Northern Hemisphere was free of ice and far warmer. North America straddled the equator and was tilted almost 50° clockwise from its present position. The ice age was likely not as sever as the Quaternary. While India was an ice box, North America was a tropical jungle. A large portion of North America was covered by low lying swamps. For the first time large land plants had covered the surface. This not only lead to the vast coal reserves of the Midwest, but changed the atmosphere. Vertebrates had not yet populated the land. There were no flowering plants. Other than that the Earth was similar to what it was before humans appeared.

Before plants moved to the land, oxygen levels in the atmosphere were less than today, but not by much. Around 350 million years ago the levels skyrocketed as plants covered the land for the first time in Earth's history. Earth currently has an atmosphere that contains about 21% oxygen. 300 million years ago, it was 35%! The highest levels ever. This not only lead to insects reaching great sizes but also greatly reduced the carbon dioxide in the atmosphere. This may have been the trigger that started the Karoo. As the south pole began to freeze, this took moisture out of the atmosphere at low latitudes, further reducing greenhouse gases and expanding the ice sheets. We do not know the extent of the glaciers at this time. If the Quaternary is a modern analog, then they may have reached from the south pole to 40° south latitude.

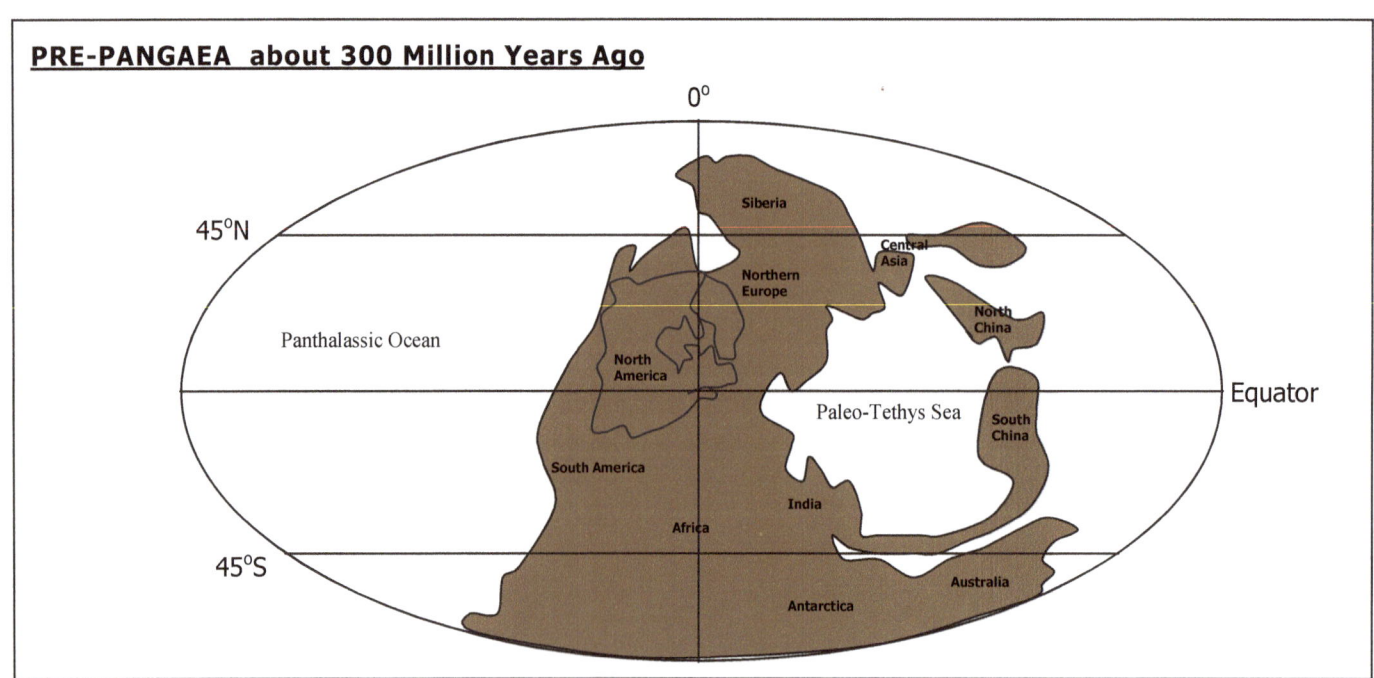

PRE-PANGAEA about 300 Million Years Ago

What Caused the Ice to Melt?

The Karoo was likely far more mild than the Quaternary. The overall elevation of the land was closer to sea level. Shallow inland seas were more widespread than today. Shallow inland seas help circulate warm air deeper inland, thus feeding the coal forests. The ice sheets likely advanced and retreated with the Milankovich Cycles as they do today. This ice age was doomed to fail from its inception. A day was shorter, about 22 hours and 10 minutes. This would have moderated day and night temperatures better than today. There were less volcanic eruptions so less dust to block out the sun. Pangaea was about to form but had not yet built the Appalachians, although the process was underway. The raising of the Appalachians when North America collided with Africa, would eventually cause the climate to become drier and shutting off new snow to the southern hemisphere. Eventually depleting the glaciers of new ice. By 260 million years ago, Earth would one again have no ice caps.

Ice Age Thoughts

I often wonder if we would even be able to recognize ancient ice ages if we were not currently living in one. There were many theories before we recognized ancient ice ages to explain the deposits we saw. Some people thought they were the products of massive landslides. Others suspected they were caused along the flanks of massive faults. Even marine origins were once popular. However, we are in an ice age. Studying these modern deposits have allowed us to recognize ancient ice deposits. It makes me wonder if our theories of other deposits for which we have no modern equivalent are correct. Such as in the case of banded iron formations.

ANDEAN-SAHARAN ICE AGE

Going back a little further, to a time before life on land at the end of the Ordovician 444 million years ago, an ice age occurred called the Andean-Saharan Ice Age. This ice age is somewhat of an enigma. Unlike most others, it was so short lived. Perhaps a maximum of one million years in duration. Although the southern polar region was likely ice covered for a period of about 30 million years (430 to 460 million years ago).

The ice age gets its name from the Andes Mountains and the Sahara Desert, where evidence of ice is prevalent. There is also evidence that Arabia and western Africa were partially covered in ice. During the late Ordovician, these areas were at or near the south pole.

What Was the Extent of the Andean-Saharan?

The extent of the Andean-Saharan was likely limited. Outside of areas known to cover ice, there is little evidence that it was widespread, and likely did not reach any further north than 50° south latitude. Some 80% of Earth's landmass was south of the equator at this time, creating a prime situation for an ice age.

Unlike today where the oceans are separated by scattered continents, there were only two oceans. There was one massive ocean more than one and a half times larger than the modern Pacific Ocean called the Panthalassic Ocean. This global ocean was directly connected to a smaller ocean the size of the Atlantic called the Iapetus or the Paleo-Tethys Ocean. The small ocean existed in an east west band. It separated what would become North America, northern Europe, and a part of Russia (connected at the time and at the equator) from the rest of the continents assembled near the south pole up to the equator. This situation essentially created one massive and global ocean current and helped bring warm air from the equator to the lower southern latitudes. This battle between growing glaciers in the south and warm air from the north restricted the ice sheets.

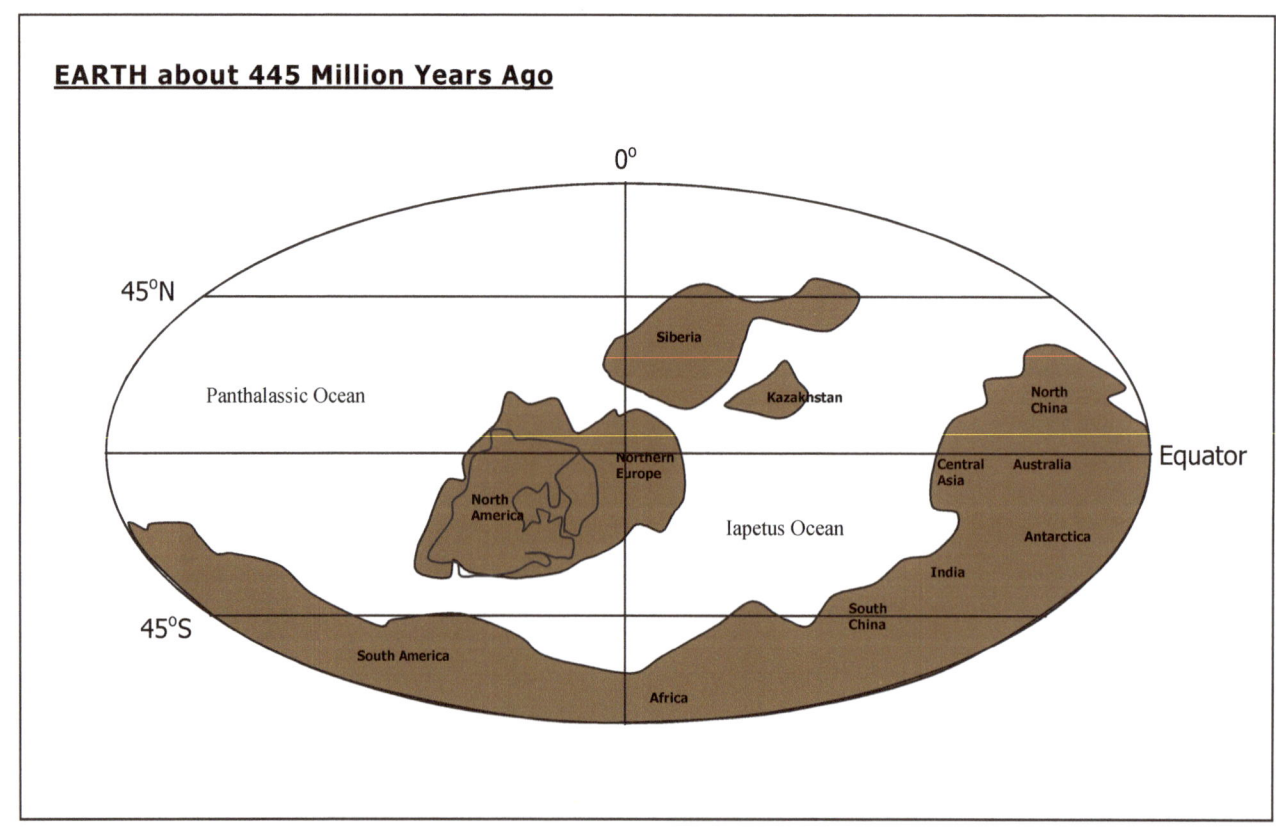

EARTH about 445 Million Years Ago

Although the ice did not reach very far north, it had a profound affect on geology. The Maquoketa Group is an assemblage of mostly shale, was deposited at the end of the Ordovician (about 445 to 450 million years ago). The Maquoketa and its equivalent units are wide spread in North America. The shale was deposited initially in deep water that progressively became more shallow. We know this by the nature of the shale. The bottom of the Maquoketa contains dark shale, typical of deeper water (likely around 400 feet deep). The top contains light colored gray and green shale, typical of shallow water (less than 200 feet deep).

During the onset of the Andean-Saharan, there was a gradual drop in global sea level due to growing ice sheets. This drop continued eventually exposing the Maquoketa Group to the surface. The Maquoketa is typically around 200 feet thick. The oceans retreated so far that about 80 to 100 feet of Maquoketa was eroded. This means that once it was exposed to the surface, the oceans withdrew another 100 feet. Since there is no evidence that the Maquoketa was uplifted through tectonic forces, so the drop in sea level had to be caused by expanding ice sheets. This global drop occurred from about 444 to 445 million years ago. By 443 million years ago all the ice had melted and the Maquoketa would again be submerged, leading to the thick limestone and dolostone deposits of the Silurian.

What Caused the Ice to Melt?

During the Ordovician life in the ocean was thriving and suffered little from the Andes-Saharan ice age. However, at the time there were no land plants or animals. The continents were barren. So there were no trees to remove carbon dioxide from the atmosphere. The estimated carbon dioxide in the atmosphere may have been 14 times modern levels. This large amount of carbon dioxide along with a global ocean current doomed this ice age from the start. It is really amazing that this ice age occurred at all.

The Andes-Saharan's short life span may have been attributed to two factors. During this time the Taconic Orogeny, a mountain building event cause by an island arc (similar to Japan) merging with the east coast of North America. This occurred from 443 to 455 million years ago. This may have been a major cause of the ice age. Once the orogeny ended, and the associated volcanic ash settled, the ice caps retreated.

There are other theories as to why this ice age occurred. They included lower solar output, a short but drastic drop in carbon dioxide levels, a temporary decrease in atmospheric density, or even a meteor impact. Although these are all viable theories, none of them can presently be tested or ruled out except of one. A meteor impact is not a likely explanation. There are only about 5 known impact craters from the end of the Ordovician. Two impacts are in Sweden, called the Lockne and Tvaren. One in Estonia called the Kardla. One in Canada called the Pilot. One last one in Illinois called the Glasford. However, the largest of these craters is only 3.7 miles in diameter (the Pilot). The smallest is the Lockne at 0.4 miles in diameter. These impacts are far too small to have affected climate on a global scale, even for the short one million year span.

CRYOGENIAN ICE AGES

There was a time before large life existed in the Oceans. If we go back to the Precambrian, we get ice age that make all later ones look like a cool summer day. During the end of the Precambrian Supereon, almost the entire Earth was frozen over when the Supercontinent Rhodinia began to break-up. Several ice ages occurred between 640 and 775 million years ago during a period called the Cryogenian. The events are often referred to as Snowball Earth Events.

What Was the Extent of the Cryogenian?

The Cryogenian ice ages occurred in at least three known intervals with one additional short interval. The youngest is the Gaskiers, which likely only lasted a couple of million years around 545 million years ago, just before large life began to dominate the oceans. The Gaskiers ice sheets likely did not cover the entire Earth. When the ice melted, life on Earth evolved very quickly in what is called the Cambrian Explosion. The Cambrian explosion marks the beginning of complex, multi-cellular life appearing in the oceans. Before this, multi-cellular life was rare and small. Most of the continents were clustered near the equator or in the Southern Hemisphere at this time in a supercontinent called Vendia. North America, Siberia, and Northern Europe were their own independent small continents at the time.

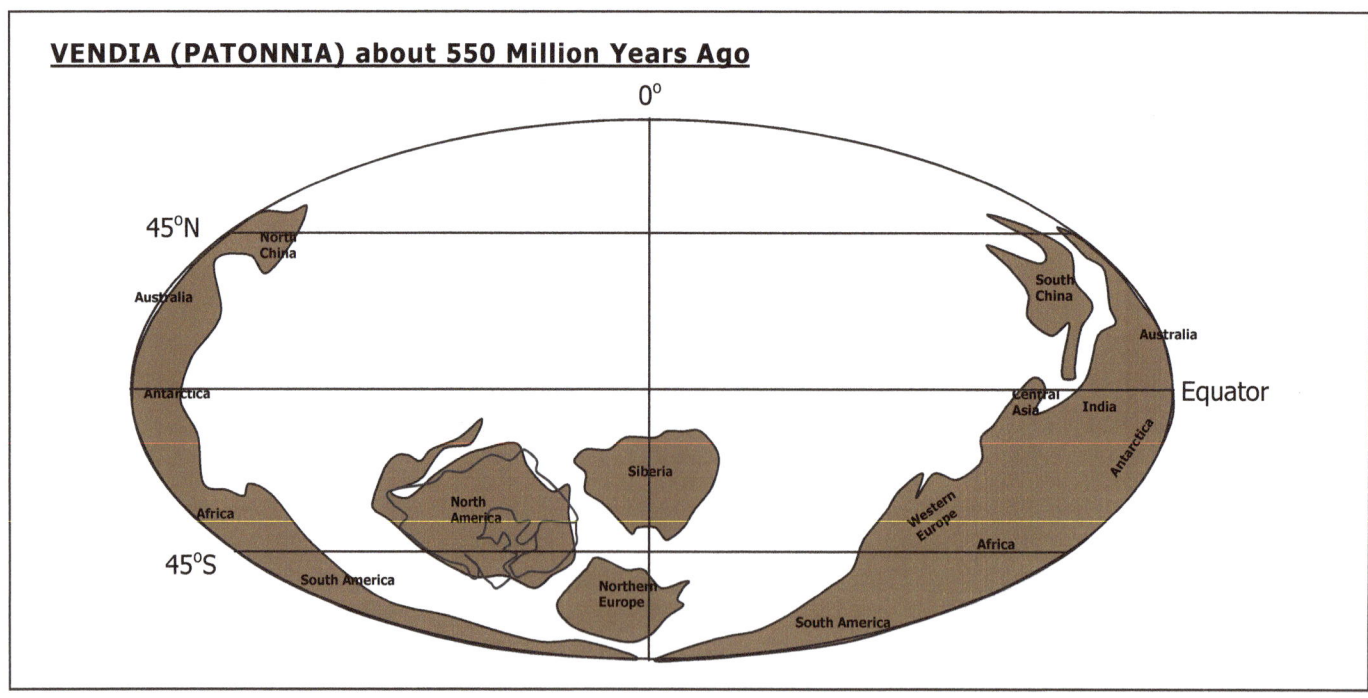

14

Banded-Iron Formation from the Upper Peninsula of Michigan. Both pieces are from the same area. The piece on the left is unaltered. The piece on the right is polished.

The red is a type of chert called jasper. The metallic gray is the mineral hematite. Hematite is used as a principle source of iron.

The Marinoan preceded the Gaskiers and lasted from 640 to 670 million years ago. If the Earth was ever totally covered by ice, this is when it happened. Multi-cellular life was about 100 million years old at the time. The continents were continuing their march away from the tropics.

The Sturtian was an ice age that existed from about 700 to 720 million years ago, just as multi-cellular life was getting started. This was the first major Snowball Earth Event. It is also the last time that a rock called "banded Iron formation" or BIF, would ever be created by the Earth. BIF is common in early Earths history from about 3.0 billion to 1.80 billion years ago, after which time they became rare. For some reason that is still undiscovered, the Earth stopped making BIFs. When the Sturtian ended, BIF would never be produced again.

The first of the Snowball Earth Events may have occurred during the Kaigas, between 730 and 775 million years ago, shortly after or during the evolution of multi-cellular life. The evidence for the Kaigas as a Snowball Earth Event is scarce, and at best, may have been a period of smaller ice ages instead of a global event. It is during this time that the tropically centered Supercontinent of Rhodinia began to break-up and most of the continents drifted towards the southern hemisphere. Rhodinia formed about 1.0 billion years ago. It's assembly stopped the Mid-continental Rift from opening up. The Mid-continental Rift formed as North America attempted to split apart where Lake Superior now stands. The rift failed, eventually closing and leading to the deposition of the richest copper deposits on Earth.

What is a Snowball Earth Event?

Snowball Earth Event is a theory first coined by Joseph Kirschvink in a short 1992 paper. These Precambrian glacial deposits were first recognized in the early 20th Century by Douglas Mawson, while studying ancient tillites in Australia. In 1998 Paul Hoffman, a Canadian born geologist, advanced the theory of Snowball Earth Events, by publishing about the cap carbonates in Namibia. Cap carbonates are thick deposits of limestone and dolostone overlying tillites. A strong indicator of rapid global warming. Vast carbonates are also found after the Ordovician glaciation in North America.

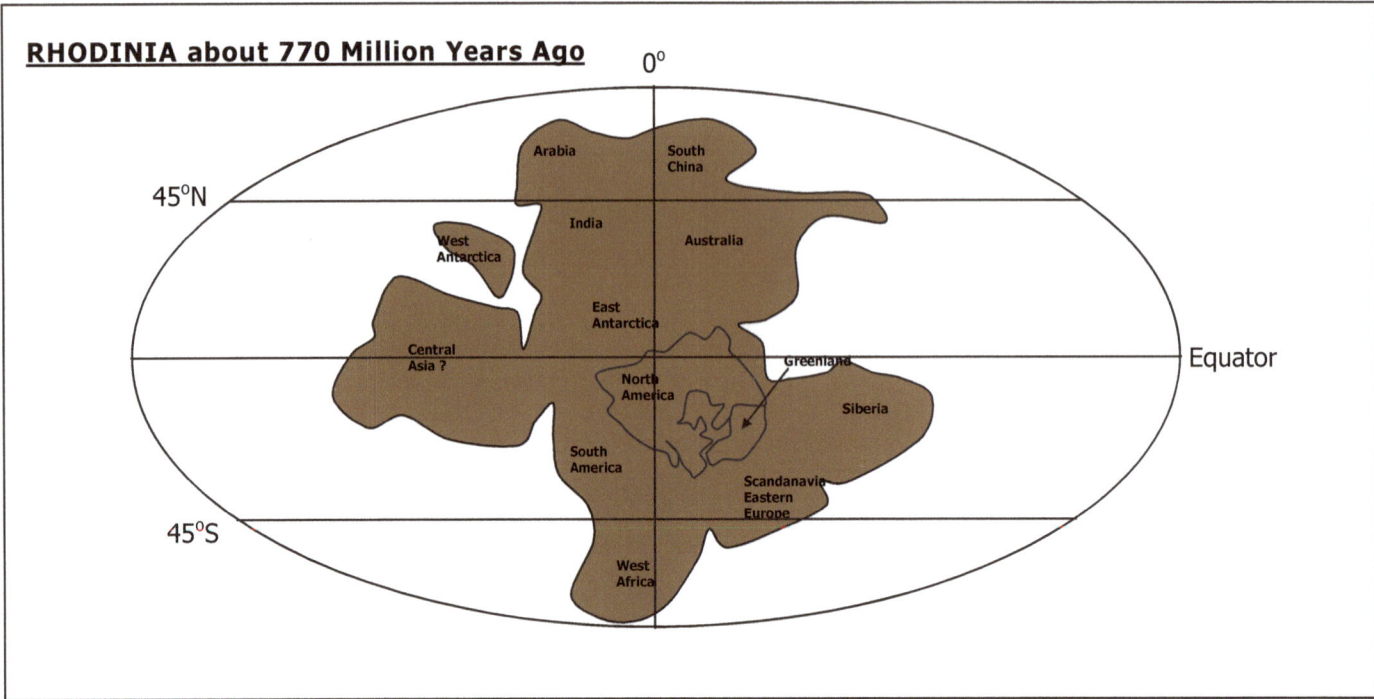

RHODINIA about 770 Million Years Ago

The Snowball Earth Events work on the assumption that the entire Earth was frozen over. Is this scenario possible? Yes. Is it likely? Doubtful. The hard geologic evidence, in the form of tillites, comes mostly from Australia, Namibia, Greenland, and Canada. The rock record elsewhere doesn't necessarily contradict such events, but it doesn't really support them either. At this time, North America was way above sea level and erosion was the dominant land shaping force.

The cap carbonates of Namibia are dozens of feet thick, but don't necessarily indicate a worldwide thawing and a massive global warming event. The Silurian deposits of North America are largely thick carbonates, some more than 800 feet thick. We know the Andean-Saharan Ice Age at the end of the Ordovician was not a global event, and the Silurian carbonates are far thicker than the older Cryogenian cap carbonates. There are many deposits around the world at the same time as the Namibia cap carbonates that do not indicate a global event. Most of the sediment deposited for a dozen million years after the Cryogenian Ice Ages are actually sandstones, especially in North America.

The reappearance of BIF has also been used as evidence. However, they are totally missing in the Marinoan glacial cycle. They also were very extensive a billion years earlier, when the Earth was much warmer. Their reappearance is likely due to a lowering of free oxygen in shallow seas, due to the Cryogenian glacial events. The Cryogenian BIFs were far limited in extent to what they were a billion years before. Most BIFs before the Cryogenian were deposited in deep ocean basins.

This isn't to say that the ice sheets were not vast. All of the land may have been frozen over. It isn't likely that the oceans were solid ice. The newly evolved multi-cellular organisms would have had a hard time surviving such a situation without a major evolutionary "make over", which is not supported by the fossil record. There is a "make over" of complex life. It just doesn't occur until after the Cryogenian ice was long gone.

We also don't have a mechanism for the initiation of such an event. The usual culprits such as, continental position, orbital changes, and life itself may have contributed to some extent. Atmospheric oxygen rose from one to five percent, right at 750 million years ago, for the first time in Earth's history, lowering carbon dioxide in the atmosphere. If the Earth was frozen solid, it had never occurred before or since. There had to be a unique combination of unidentified factors at play that we do not yet understand. That's okay. We don't have all the answers. It's what keeps science interesting.

What Caused the Ice to Melt?

There is no doubt that the Cryogenian Ice Ages were vast, cold, and brutal. Even if the Earth wasn't frozen solid, it would still have been a frozen hell. The break-up of Rhodinia would have caused massive volcanic eruptions, pumping a vast amount of heat capturing carbon dioxide into the atmosphere. Bacteria and methane from the ocean bottom may also have helped melt the ice.

Ice Age Thoughts

The study of ice ages isn't just a geologic endeavor. It involves chemists, physicists, astronomers, archeologists, and the public. It is important for people to understand how our Earth works. Teachers provide the basic framework for minds to ponder how the Earth that we see today came to be. It is equally important for students to realize that in order to understand the present and plan for the future, we need to understand the very distant past.

HURONIAN ICE AGES

In Canada, along the northern part of Lake Superior lay some very old rocks that belong to the Huronian Supergroup. The Huronian Supergroup is an assemblage of rocks that contain volcanics at the base and alternating sedimentary rocks throughout. In places the Huronian rocks are nearly five miles thick! That's a lot of geologic information. These rocks record the break-up of the supercontinent Kenorland. Kenorland included not only most of Canada but also Wyoming, parts of Montana and Idaho, Antartica, along with Western Australia and Northern Europe. Kenorland was a supercontinent for its day but it was small. The total continental mass of the Earth was about 30% to 40% what it is today. This supercontinent is more difficult to decipher than the much younger Rhodinia because it existed over one and a half billion years earlier (2.45 to 2.60 billion years ago).

The further back in time you go, the more incomplete the overall rock record is. This is because the Earth is a very dynamic planet. The continents have been in motion for at least three billion years. Dozens of mountain ranges have been raised and eroded. Entire areas the size of small continents have been added to other continents or completely eroded. This makes deciphering the distant past difficult, but not impossible. The exact placement of Kenorland is not known, although it was likely centered entirely in the northern or southern hemisphere. India was not a part of Kenorland and likely was its own continent near the north or south pole.

The Huronian deposits and equivalent units, preserve the first complete record of a successful splitting of continents. The supercontinent of Kenorland broke apart from about 2.20 to 2.45 years ago. This is also the first time in the rock record that red beds appear. Red beds are sediments that are "rusted" a red color. Red beds won't form without free oxygen in the atmosphere. The event at which Earth's atmosphere began to contain free oxygen is called the "Great Oxygenation Event". Life had been around on the single celled level since about 3.9 billion years ago but it wasn't until about 2.35 billion years ago that the Great Oxygenation Event occurred. It began to change the atmosphere, even though life would be restricted to the oceans for another two billion years.

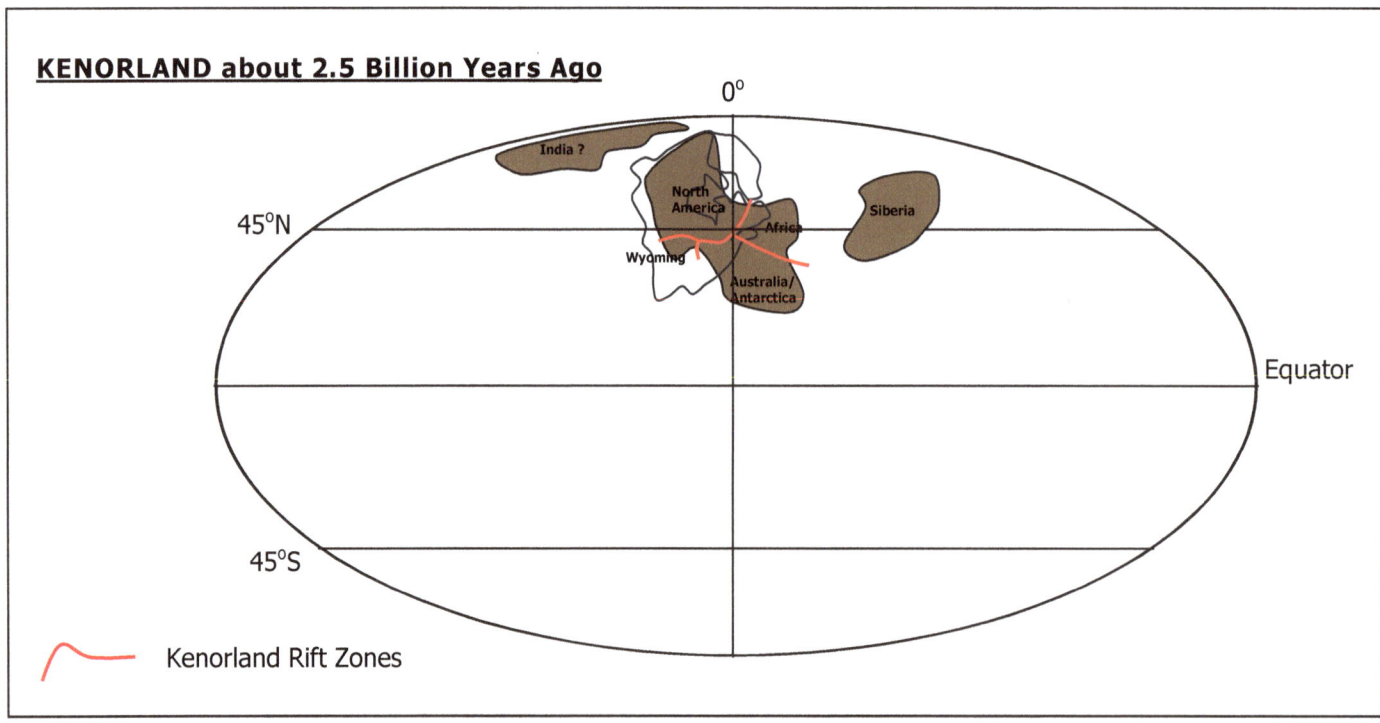

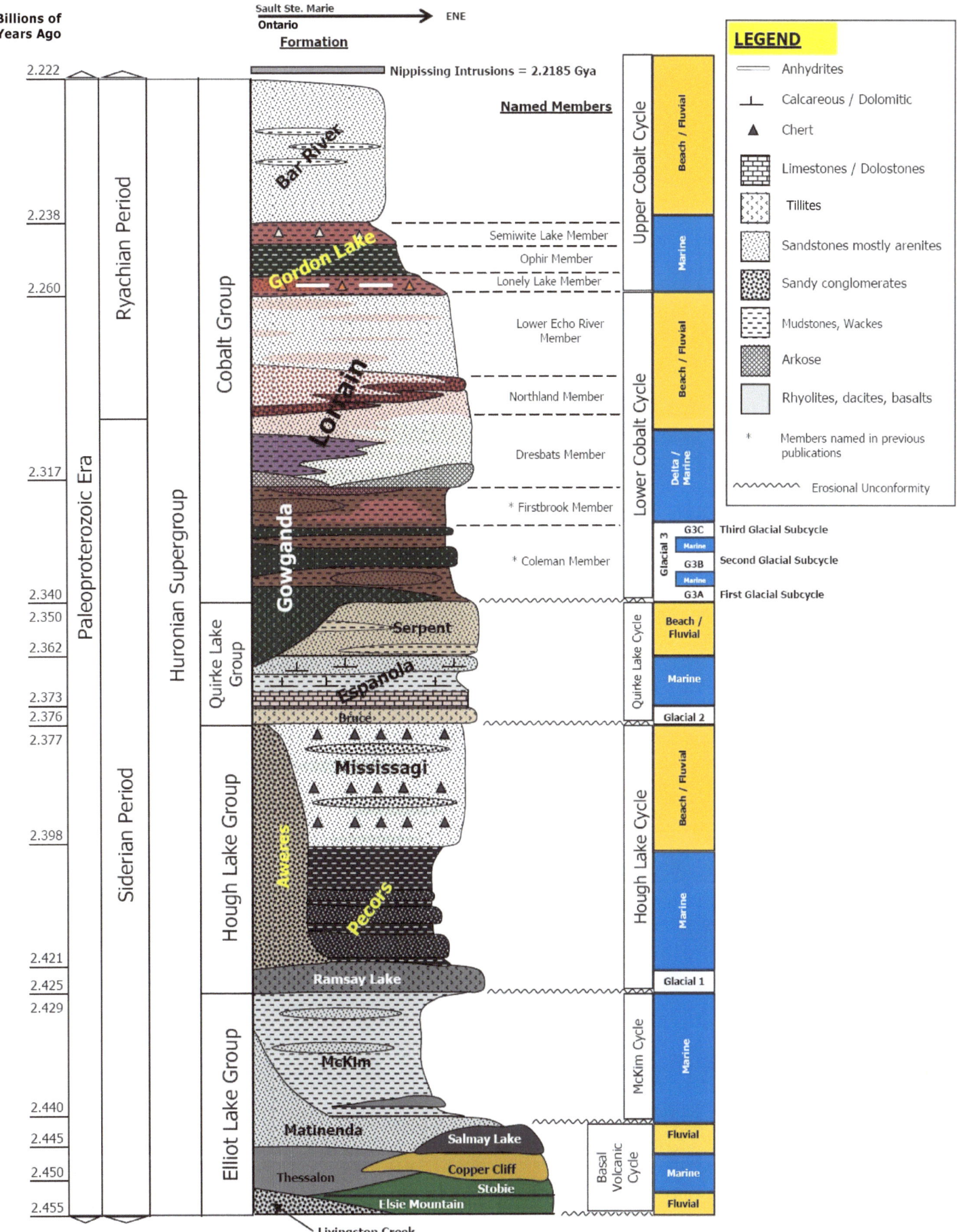

This is a stratigraphic column of the rocks within the Huronian Supergroup. It shows all of the main rock types within the Huronian and their depositional environments. All three glacial cycles are labeled. The first occurring with the deposition of the Ramsay Lake Formation. The second during the deposition of the Bruce Formation. The third during the Gowganda Formation, which contains at least three subcycles.

What was the Extent of the Huronian Ice Ages?

This change in the atmosphere and oceans lowered global temperatures and the ice moved in. Ontario, the Upper Peninsula of Michigan, and Wyoming record at least three glacial cycles. The oldest was about 2.425 billion years ago, after the supercontinent of Kenorland had been breaking up. It's duration was only about four million years. The fact that similarly aged deposits in the Northwest Territories of Canada and in Finland do not contain tillites, suggests it was a short lived ice age with a small extent, perhaps similar to the Quaternary Ice Ages.

The second ice age in the Huronian, occurred from about 2.373 to 2.376 billion years ago. This one was even shorter than the first with about the same extent. During this time the break-up of Kenorland was fully underway.

The youngest of the Huronian glaciations occurred from about 2.325 to 2.340 billion years ago. This is significantly longer than the first two. In Ontario, there are at least three identifiable subcycles. Subcycles are minor glacial advances and retreats that occur within a glacial cycle. The extent of the ice was also greater. The ice covered the same area as the first two but also included the Northwest Territories, Finland, and parts of Australia. Some argue that this was a Snowball Earth situation. This isn't likely. There are still areas without any tillites, indicating that the coverage was not global. Also cap carbonates have been used as an argument for a Snowball Earth during the Cryogenian how about in the Huronian? No. Other than a couple of exceptions, there are almost no carbonates after each of the Huronian glacial cycles. Most of the deposits after each of the three cycles are followed by thick sandstone and shale deposits, not carbonates (the Espanola Formation is the exception). The short span of most ice sheets indicate that the glaciers had moved to lower latitudes but likely stayed away from the equator. Earth may have had larger ice caps for longer at both poles than today, during the time in between the three Huronian Ice Ages. This time may have been the longest period in the history of our planet where ice caps at the poles persisted from about 2.20 to 2.43 billion years ago. A span of 230 million years!

Tillite present in the Gowganda Formation of the Huronian Supergroup. This photo is off Ontario Route 638, near Rock Lake.

Notice the red clasts (Older Arcean Granite) "floating" in the gray fine grained matrix (originally deposited as clay and silt). These red pebbles to cobbles were transported by the ancient continental glacier and incorporated into the diamicton that would later harden into tillite.

The Huronian ice ages are my favorite. The evidence for glaciation combined with the break-up of Kenorland provides a unique way to study how different processes of the Earth interact. In the same areas you have intense volcanism followed by quiet sedimentary deposits that were occasionally intruded by glaciers. There is no modern analogy for this situation. Earth was also a very different planet than today. You would not have recognized it. The atmosphere was less than 0.1% oxygen. Methane, sulfur dioxide, nitrogen, and carbon dioxide made up almost all of the atmosphere. The sky would have been a yellowish color and the oceans green. Life was around, but was only single celled and restricted to the oceans. Although life was abundant in the oceans, the land was a totally different situation. There were no plants, no animals, no summer night sounds of crickets, nor was there the smell of

flowers in the air. The land was dead, not even the smallest insect. There was no ozone to protect the Earth from ultraviolet radiation. The moon was much closer. The days, much shorter. The sun was only about 85% as bright as it is today. A day was less than 20 hours long. Earth was 2.5% closer to the sun than it is today. Earth certainly has changed!

What Caused the Ice to Melt?

When Kenorland was fully assembled about 2.5 billion years ago, continental arrangement was the opposite of today. The majority of land was either in the southern or northern hemisphere, with India likely a continent at the North Pole. The break-up of Kenorland started by splitting Ontario and the Upper Peninsula from Wyoming. This process was almost complete around 2.30 billion years ago. Other rifts formed elsewhere from about 2.20 to 2.30 billion years ago. By 2.2 billion years ago Kenorland was gone. Also, additional volcanic activity occurred after the deposition of the Huronian Supergroup from 2.0 to 2.2 billion years ago. The most extensive of the post Huronian volcanics are the Nippissing Intrusions. The exact cause of this renewed post rifting volcanism is unclear, but was likely a major cause in melting the ice. The later volcanism may have been the last "gasps" of a dying pre-plate tectonics mechanism. Plate tectonics was fully underway by two billion years ago and hasn't changed much since. However, from about two to three billion years ago the Earth was in transition from a possible Venus style volcanism to plate tectonics.

After the Huronian glaciers retreated, Earth appears to have remained ice free for more than 1.55 billion years! This is an immense amount of time. It is almost three times longer than complex life has been around on Earth. It shows how the our planet of today is not the planet of yesterday.

COMPARATIVE PRIMEVAL and MODERN ATMOSHERES						
Planet	Pressure (Bars)	Carbon Dioxide (%)	Nitrogen (%)	Argon (%)	Oxygen (%)	Methane and Sulfur Dioxide (%)
Venus (Today)	92	96.5	3.5	0.00007	Trace	Trace
Earth (Today)	1.013	0.033	78	0.01	21	Trace
Earth (2.5 Billion Years Ago)	0.5 ?	90	5	≤ 1	< 1	3

Earth's ancient atmospheric pressure is conjectural at best.

OLDER ICE AGES

Based on deposits of tillite in South Africa within the Delfkom Formation, it is believed that there was a restricted ice age around 2.9 billion years ago called the Pangola. During this time, modern plate tectonics was just beginning and was not like it is today. All of the Earth's landmasses are suspected to exist in a small continent about the size of Australia called "Ur", with small volcanic islands and masses of granite spread across the globe. Ur consisted of small parts of modern day India, Africa, Australia, and Antarctica. What is now North and South America, Asia, and Europe, would not begin form for another 300 million years. The total continental land mass was probably 10% to 15% what it is today.

Over the countless eons, the landmasses of Earth have been getting gradually larger as they shift around the globe. This is due to lighter rocks like granite moving upwards from deep within the Earth over billions of years. As a result, the oceans still have the same amount of water, they are just deeper than they once were, because of the cooling of the Earth's crust over billions of years.

Ur was likely located in the middle to near polar latitudes of the Southern Hemisphere. The tillites present are the only known ones on Earth from this time. It is unclear if these tillites were deposited by glaciers or as debris flows from glaciers at high altitudes.

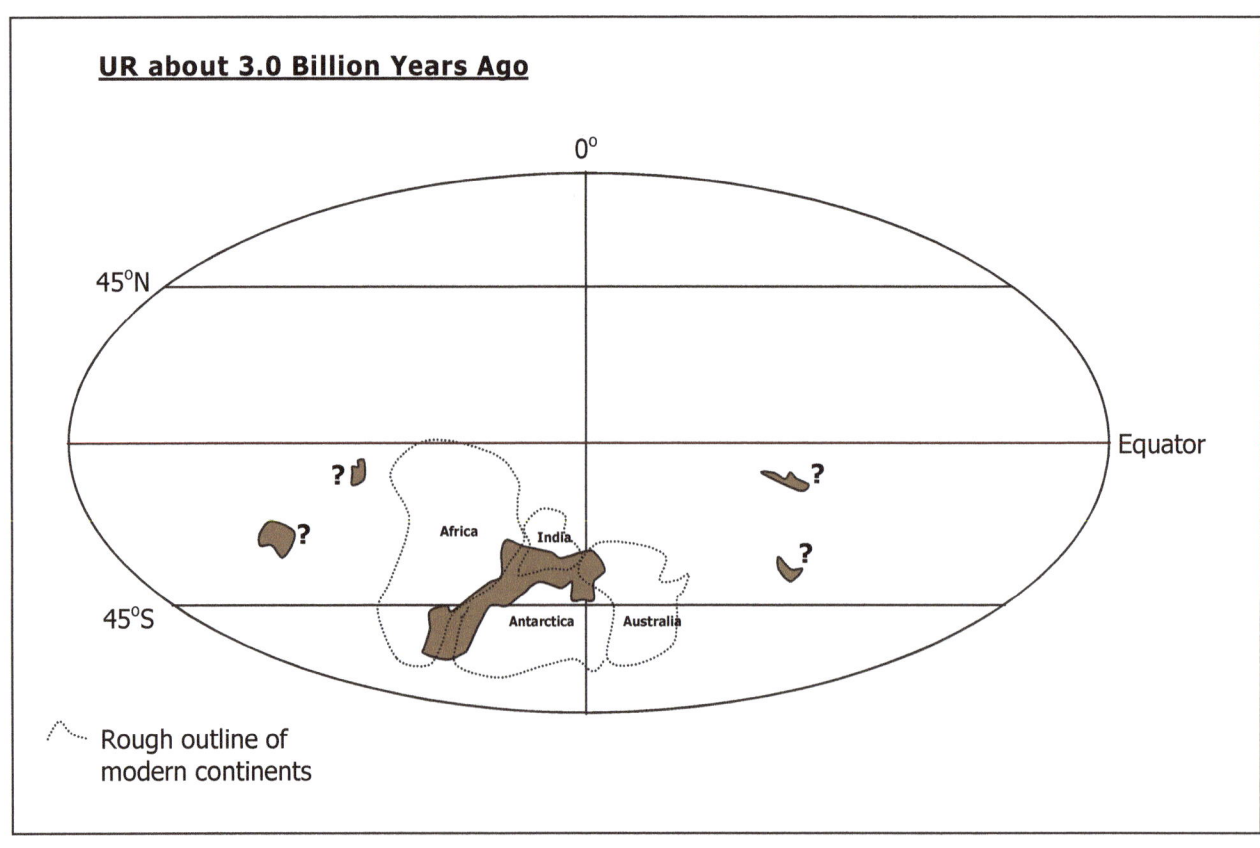

Before 2.9 billion years ago, there is no evidence of glaciers existing on Earth, nor is there likely to be any discovered. From the formation of Earth, some 4.6 billion years ago to 2.9 billion years ago, our planet was not hospitable. Although single celled ocean life appears around 3.9 billion years ago, Earth was not a friendly place. The moon was almost 30% closer to the Earth than it is today. This would have caused tides and waves much larger than anything seen today. There was no oxygen in the atmosphere. Acid rain was far more common. The Earth was just getting over being struck by asteroids a half a mile across or larger at a rate of one per century. Not the life giving planet of today.

Ice Age Thoughts

The vast ice sheets that covered North America picked-up all sorts of things as they moved across the landscape. Large glacial boulders, or erratics, can travel hundreds of miles. The largest one in Illinois is a boulder of granite 22 by 10 by 11 feet in size, weighing near 100 tons! It was discovered in Southern Illinois, only 50 miles from the furthest extent of the Quaternary ice sheets. It likely originated within 50 miles of Lake Huron in Ontario (north east of Sault Ste. Marie). It has been dated at 2.66 billion years and traveled about 700 miles! You can find out more information on the Illinois State Geological survey's website:

www.isgs.illinois.edu/southern-illinois-mega-erratic-revisited-266-billion-year-old-find

Other rare occurrences of precious stones and minerals have also been found in glacial deposits. Gold, diamonds, and copper have all been found dozens to hundreds of miles from their places of origin.

WHY ARE ICE AGES IMPORTANT?

The study of our planet's past is the key to figuring out its future. The record of Earth's evolution is in the rocks. The further back you go, the less evidence there is. Understanding ice ages that occurred millions or even billions of years ago, is still relevant. We are currently still in an ice age. Will the Earth get colder or warmer? For how long? In order to answer these questions we need to look back as far as we can. All knowledge is useful, even if it seems arbitrary at the time. Not knowing is the greatest risk of all. Ignorance may be bliss, but it is also deadly.

Will the Earth continue to warm or get colder? Based on what we know about the past glacial events of the 2.5 million years and the ones extending back 2.5 billion years, the long term answer is the Earth will ultimately warm and the poles will eventually melt. As for the next few thousand years, the Earth will likely continue to warm. However, if past ice ages are a clue as to future ones, we have at least one more major ice advance in our future. With our participation in altering the climate, the timing of the next ice age is difficult to compute. We are presently about midway through a typical interglacial period. The logical assumption would be the next ice age should begin around 12,000 AD. Our presence on this planet has changed things. That could be pushed back or moved forward. In the very short term, 1000 years or so, the Earth will likely warm not cool.

So. What can be done? We need to let the Earth cycle naturally. If it gets warmer or colder, we need to be able to adapt either way. Our ancestors 50,000 years ago did it. Why can't we? We need to stop fighting nature and build with it. We need to implement smarter more sustainable technology. Stop chasing money. Live for the future, not the right now. We need to greatly reduce our population and our consumption. By doing these things, we can more readily adapt to change instead of reacting to it. We can either ease into new era of human civilization or quickly burn a horrible death in this one. It's up to us.

ICE AGE PHOTO GALLERY

Illinois Age glacial sand and gravel outwash of the Pearl Formation, Central Illinois. U.S. dollar coin for scale.

Wisconsin glacial erratic (about 5 feet wide) in the Vermilion River, Matthiessen State Park, Illinois.

Diamicton of the Tiskilwa Formation, north-central Illinois. Outcrop is about 8 feet high.

ICE AGE PHOTO GALLERY

Geomorphic feature in Cook County, Illinois at Swallow Cliffs. This is a glacial end moraine.

This hill was cut out when glacial melt water eroded it from the left to right side of the photo.

Cypress trees in Southern Illinois . These trees exist in Horseshoe Lake. Horseshoe Lake was a meander in the Mississippi River, until it was cut-off at the end of the last ice age.

ICE AGE PHOTO GALLERY

The mouth of Dead River in Illinois State Beach State Park. The Dead River is the only River in Illinois that still flows into Lake Michigan (background). The Dead River cuts through the sand dunes along Lake Michigan that began to form after the last ice age. Backpack is 18 inches tall.

A piece of the Gowganda Tillite . This is a glacial erratic that is on display outside the Illinois State Geological Survey. This erratic was found about 600 miles from its place of origin in Ontario. The Gowganda Formation was tillite deposited 2.34 billion years ago. It was reincorporated as a glacial erratic in the most recent ice age. My how geologic processes repeat themselves.

REFERENCES

Aspler, L.B., 1998, *Two Neoarchean Supercontinents? Evidence from the Paleoproterozoic*, Journal of Sedimentary Geology, 120(1998)75-104

Baumann, S.D.J., 2011, *Preliminary Redefinition of the Cobalt Group in the Southern Geologic Province, Ontario, Canada*, Midwest Institute of Geosciences and Engineering, G-012011-2A

Dahl, P.S., 2007, *2480 Ma Mafic Magmatism in the Northern Black Hills, South Dakota: A New Link Connecting the Wyoming and Superior Cratons*, Large Igneous Provinces Commission

Ernst, R.E., 2009, *Reconstructing Ancient Continents Using the Large Igneous Province Record: Implications for Mineral, Hydrocarbon, and Earth Systems*, Large Igneous Provinces Commission

Hansel, A.K., 1996, *Wedron and Mason Groups: Lithostratigraphic Reclassification of Deposits of the Wisconsin Episode, Lake Michigan Lobe Area*, Illinois State Geological Survey, Bulletin 104

Killey, M.M., 2007, *Illinois' Ice Age Legacy*, Illinois State Geological Survey, Geoscience Education Series 14

Mojzsis, S.J., 2012?, *Life and the Evolution of Earth's Atmosphere*, Department of Geological Sciences, University of Colorado

Muller, R.A., 2002, *Ice Ages Astronomical Causes: Data, Spectral Analysis, and Mechanisms*, Springer Praxis Books

Raub, T.D., 2007, *A Pan-Precambrian Link Between Deglaciation and Environmental Oxidation*, 10th International Symposium on Antarctic Earth Sciences

Rovey, C.W., 2011, *Summary of Early and Middle Pleistocene Glaciations in Northern Missouri, U.S.A.*, Quaternary Glaciations-Extent and Chronology, Chapter 43, pp. 553-561

Som, S.M., 2012, *Air Density 2.7 Billion Years Ago Limited to Less Than Twice Modern Levels by Fossil Raindrop Imprints*, Nature, Vol. 484, pp. 359-362

MORE INFORMATION

You can find out more free scientific information on ice ages at the following websites:

Institute on Lake Superior Geology
www.lakesuperiorgeology.org

Illinois State Geological Survey
www.isgs.illinois.edu

Midwest Institute on Geosciences and Engineering
www.mige-web.org

Ontario Geological Survey
www.geologyontario.mndm.gov.on.ca

Wisconsin Geological and Natural History Survey
Www.wgnhs.uwex.edu/

www.ingramcontent.com/pod-product-compliance
Lightning Source LLC
Chambersburg PA
CBHW050407180526
45159CB00005B/2185